Mathematics WARM-UPS

Grade 8

STATE COMMON CORE STANDARDS

WALCH EDUCATION

1 2 3 4 5 6 7 8 9 10
ISBN 978-0-8251-7149-9

J. Weston Walch, Publisher
Portland, ME 04103
www.walch.com
Printed in the United States of America

Table of Contents

Introduction

Mathematics Warm-Ups for Common Core State Standards, Grade 8 is organized into five sections, composed of the domains for Grade 8 as designated by the Common Core State Standards Initiative. Each warm-up addresses at least one of the standards within the following domains:

- The Number System
- Expressions and Equations
- Functions
- Geometry
- Statistics and Probability

The Common Core Mathematical Practices standards are another focus of the warm-ups. All of the problems require students to "make sense of problems and persevere in solving them," "reason abstractly and quantitatively," and "attend to precision." Students must "look for and make use of structure" when graphing functions and interpreting exponents. Students have opportunities to "use appropriate tools strategically" when they use graph paper to explore transformations. A full description of these standards can be found at http://www.walch.com/CCSS/00006.

The warm-ups are organized by domains rather than by level of difficulty. Use your judgment to select appropriate problems for your curriculum.* The problems are not necessarily meant to be completed in consecutive order—some are stand-alone, some can launch a topic, some can be used as journal prompts, and some refresh students' skills and concepts. All are meant to enhance and complement your Grade 8 mathematics program. They do so by providing resources for those short, 5- to 15-minute interims when class time might otherwise go unused.

* You may select warm-ups based on particular standards using the Standards Correlations table.

About the CD-ROM

Mathematics Warm-Ups for Common Core State Standards, Grade 8 is provided in two convenient formats: an easy-to-use reproducible book and a ready-to-print PDF on a companion CD-ROM. You can photocopy or print activities as needed, or project them on a screen via your computer.

The depth and breadth of the collection give you the opportunity to choose specific skills and concepts that correspond to your curriculum and instruction. Use the table of contents and the standards correlations to help you select appropriate tasks.

Suggestions for use:

- Choose an activity to project or print out and assign.
- Select a series of activities. Print the selection to create practice packets for learners who need help with specific skills or concepts.

Standards Correlations

Mathematics Warm-Ups for Common Core State Standards, Grade 8 is correlated to five domains of CCSS Grade 8 mathematics. The page numbers, titles, and standard numbers are included in the table that follows. The full text of the CCSS mathematics standards for Grade 8 can be found in the Common Core State Standards PDF at http://www.walch.com/CCSS/00001.

Page number	Title	CCSS addressed
The Number System		
1	Irrational Numbers	8.NS.1
2	Where Do They Go?	8.NS.2
Expressions and Equations		
3	Tearing and Stacking Paper	8.EE.4
4	Linear vs. Nonlinear	8.EE.5
5	Graphing Linear Functions	8.EE.5
6	The Seesaw Problem	8.EE.7b
7	Systems of Linear Equations	8.EE.8a, 8.EE.8b
8	Exploring Systems of Equations	8.EE.8b
9	Nathan's Number Puzzles I	8.EE.8b
10	Understanding Systems of Linear Equations	8.EE.8b
11	Nathan's Number Puzzles II	8.EE.8c
Functions		
12	Contaminated Drinking Water	8.F.1
13	Cars and Drivers	8.F.2
14	Slope-Intercept Form	8.F.3
15	Slope-Intercept Equations	8.F.3

(*continued*)

Page number	Title	CCSS addressed
16	How Long? How Far?	8.F.4
17	Game Time	8.F.4
18	Hedwig's Hexagons	8.F.4
19	Finding Slope-Intercept Equations	8.F.4
20	Up and Down the Line I	8.F.4
21	Up and Down the Line II	8.F.4
22	Butler Bake Sale	8.F.4
23	Looking for Lines	8.F.4
24	Thinking About Equations of Linear Models	8.F.4
25	Thinking About Change and Intercepts in Linear Models	8.F.4
26	Freezing or Boiling?	8.F.4
27	Border Tiles for Goldfish Pools	8.F.4
28	Table Patterns	8.F.4
29	Parachuting Down	8.F.4
30	Sketching a Graph	8.F.5
31	Interpreting Graphs I	8.F.5
32	Interpreting Graphs II	8.F.5
33	Ferris Wheel Graphs	8.F.5
34	Roller Coaster Ride	8.F.5
35	Graphing People Over Time	8.F.5
36	Renting Canoes	8.F.5
37	Grow Baby!	8.F.5
Geometry		
38	Double Trouble	8.G.2
39	Up, Down, and All Around	8.G.3

(*continued*)

Page number	Title	CCSS addressed
40	Translate This!	8.G.3
41	Mirror, Mirror on the Wall	8.G.4
42	Copy Cat	8.G.4
43	Tree Dilemma	8.G.4
44	Road Trip	8.G.7
45	Comparing Cylinders	8.G.9
46	Concession Concern	8.G.9
47	Popcorn Pricing	8.G.9
48	Juice Packaging	8.G.9
49	Cylinder Expressions	8.G.9
50	Paper Cylinders	8.G.9
51	Tennis Ball Packaging	8.G.9
Statistics and Probability		
52	Analyzing Scatter Plots	8.SP.1
53	Mr. Wiley's Baby	8.SP.2

NAME:

THE NUMBER SYSTEM
CCSS 8.NS.1

Irrational Numbers

Use the information below and what you know about irrational numbers to answer the question that follows.

> An irrational number is a number that cannot be expressed as a fraction. Any decimals that are not terminating and do not repeat are irrational numbers. More technically, a rational number is a number that can be expressed in the form $\frac{x}{y}$, where x and y are integers and y is not 0.

Is $\sqrt{2}$ an irrational number? Why or why not? Explain your thinking.

Where Do They Go?

Using the number line below, show approximately where each number would fall. Explain your thinking.

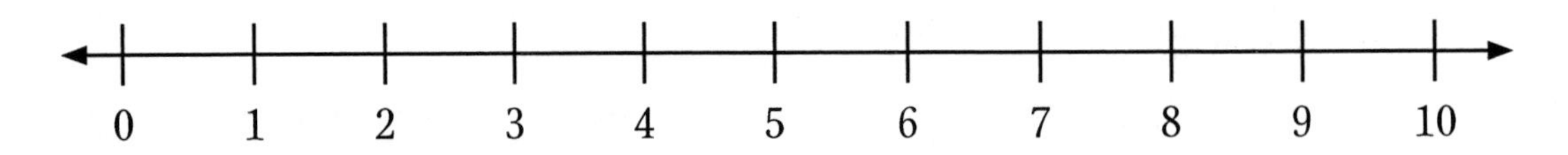

1. $\sqrt{96}$

2. $\sqrt{35}$

3. $\sqrt{24}$

4. $\sqrt{17}$

NAME:

EXPRESSIONS AND EQUATIONS

CCSS 8.EE.4

Tearing and Stacking Paper

Read the scenario that follows, and then answer the questions.

> Mr. Andres poses the following problem to his math class: Take a large sheet of paper and tear it exactly in half. Then you have 2 sheets of paper. Put those 2 sheets together and tear them exactly in half. Then you have 4 sheets of paper. Continue this process of tearing and putting together for a total of 50 tears.

If the paper is only $\frac{1}{1,000}$ of an inch thick, how many sheets of paper would there be?

How thick or tall would the stack of paper be?

NAME:

EXPRESSIONS AND EQUATIONS

CCSS 8.EE.5

Linear vs. Nonlinear

A linear equation is an equation that can be graphed by a straight line. A nonlinear equation is an equation that cannot be represented by a line. Determine whether the following graphs are linear or nonlinear. Write your answer on the line below each graph.

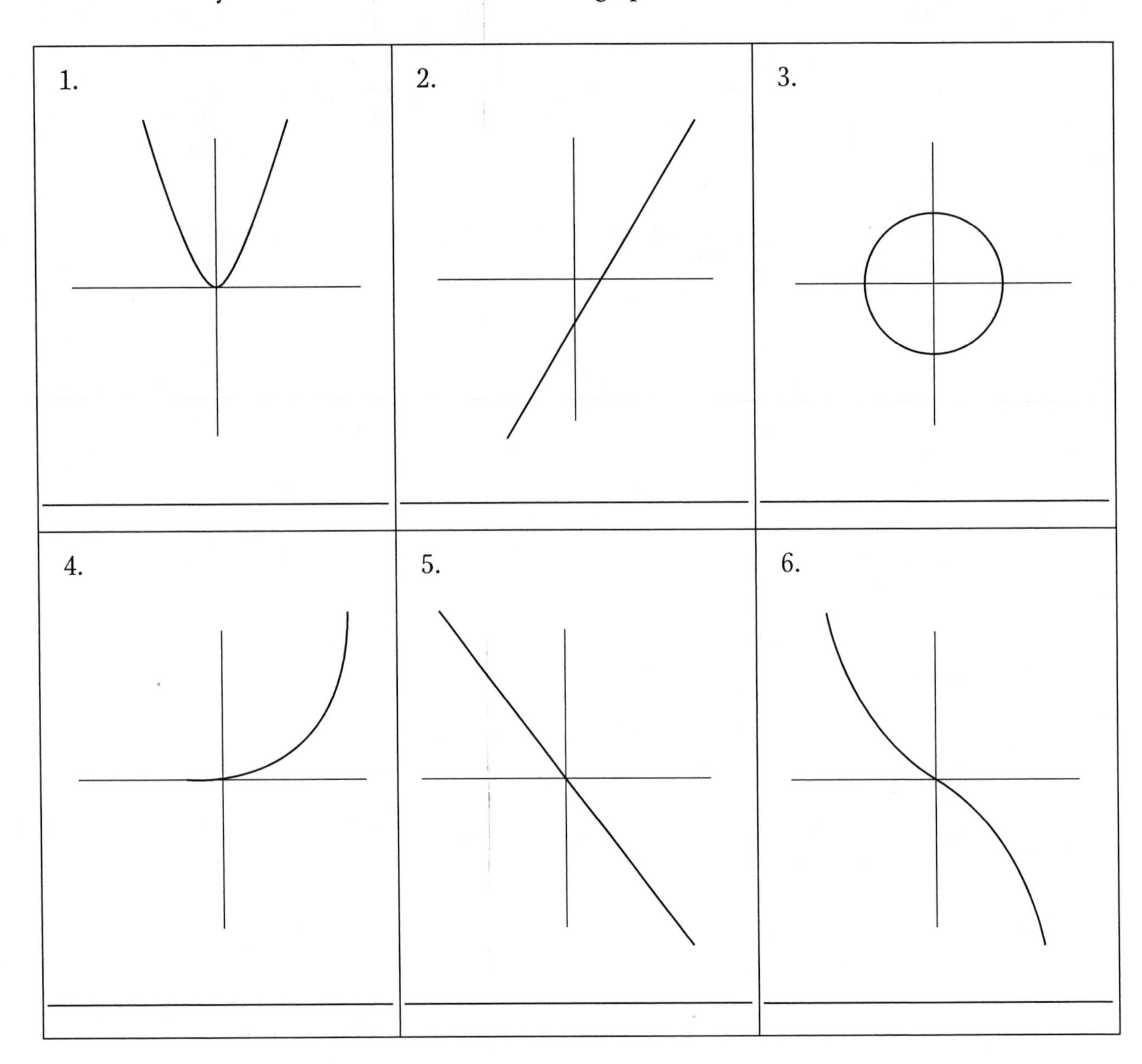

NAME:

EXPRESSIONS AND EQUATIONS
CCSS 8.EE.5

Graphing Linear Functions

Using a function to generate output will lead to the production of a set of ordered pairs. We use ordered pairs when plotting points on a coordinate plane. For example, look at the chart below. The output can be used as the y-coordinate.

Input (x)	Function ($2x + 1$)	Output (y)
1	$2(1) + 1$	3
2	$2(2) + 1$	5
3	$2(3) + 1$	7
4	$2(4) + 1$	9

The ordered pairs created are (1, 3), (2, 5), (3, 7), and (4, 9). They all lie on a straight line. You can connect the points to see all the other points on the line.

Graph the line generated by each function below. Use x-values from at least 0 to 5.

1. $y = 2x - 2$

2. $y = x + 1$

The Seesaw Problem

Read the information in each problem, and then answer the questions.

1. Caleb and his friend Alex are playing on the seesaw at the playground. Alex weighs 70 pounds. Caleb and Alex balance perfectly when Alex sits about 3 feet from the center and Caleb sits about $2\frac{1}{2}$ feet from the center. About how much does Caleb weigh?

2. Alex's identical twin brother Samuel joins them and sits next to Alex. Can Caleb balance the seesaw with both Alex and Samuel on one side, if Samuel weighs about the same as Alex? If so, where should Caleb sit? If not, why not?

NAME:

EXPRESSIONS AND EQUATIONS

CCSS 8.EE.8a, 8.EE.8b

Systems of Linear Equations

Taber is learning about systems of linear equations. He has come up with some questions regarding equations such as $y = 3x + 8$ and $y = -5x + 11$. Help him by answering the questions below.

1. What is the objective for solving a system of equations such as the one given?

2. How could you find the solution by graphing?

3. How could you find the solution by using tables of values?

4. How might you find the solution for the system without a graph or a table?

5. What would you look for in the table, the graph, or in the equations themselves that would indicate that a system of equations has no solution?

Exploring Systems of Equations

Scientists attempt to model our world with equations. In many cases, a situation can only be modeled using multiple equations. This is called a system. Systems involve many variables that influence other variables. What systems can you name? Solving a system of equations is sometimes called solving simultaneous equations.

1. a. Graph the following equation on graph paper.

   $y = 2x - 3$

   b. Fill in the table below.

x	0	1	2	3	4	5
y						

2. a. Now, graph the following equation on the same set of axes as your first equation.

   $y = -3x + 12$

   b. Fill in the table below.

x	0	1	2	3	4	5
y						

3. In the tables, which x- and y-coordinates are the same?

4. Looking at your graph, where do the lines intersect?

5. Substitute the point of intersection into the first and second equations. What do you notice?

6. Now choose the first point in the first table to substitute into both equations. What do you notice?

7. What do you think would happen if you substituted any other point besides the point of intersection into both equations?

Nathan's Number Puzzles I

Nathan has written pairs of linear sentences in symbols. Now he's trying to create word puzzles to go with them. Help him by creating a word puzzle for each pair of linear sentences below. Then find the numbers that fit his equations.

1. $x + 3y = 5$

   $3x + y = 7$

2. $x + 3y = 10$

   $3x + 14 = 2y$

3. $x + 2y = 7$

   $x + 2y = 17$

Understanding Systems of Linear Equations

Marika has asked you to help her understand how to solve systems of equations. Solve each system of equations below using a different strategy. Then explain to Marika why you chose that strategy for that system. Which are best solved by substitution? Which might be easily graphed? Which could be solved by elimination?

1. $y = x - 1$

   $3x - 4y = 8$

2. $3x + 2y = -10$

   $2x + 3y = 0$

3. $x + y = -10$

   $0.5x + 1.5y = 5$

4. $3x - 2y = 6$

   $-2x + 3y = 0$

Nathan's Number Puzzles II

Nathan likes to make up puzzles about integers. Some of his recent puzzles are below. Write symbolic sentences that represent Nathan's puzzles. Then solve each puzzle.

1. Two numbers have a sum of 10. If you add the first number to twice the second number, the result is 8. What are the numbers?

2. One number is twice as large as a second number. The sum of the two numbers is 15. What are the numbers?

3. The first number minus the second number is 2. Twice the first number minus twice the second number is 4. What are the numbers?

Contaminated Drinking Water

A pond used for drinking water has been contaminated by a recent forest fire, and has to be drained. The pond contains approximately 29,600 gallons of water and will be drained at a rate of 925 gallons per hour.

1. How many gallons will remain in the pond after 5 hours of draining? After 10 hours? After 15 hours?

2. Graph each input and its corresponding output as a set of ordered pairs.

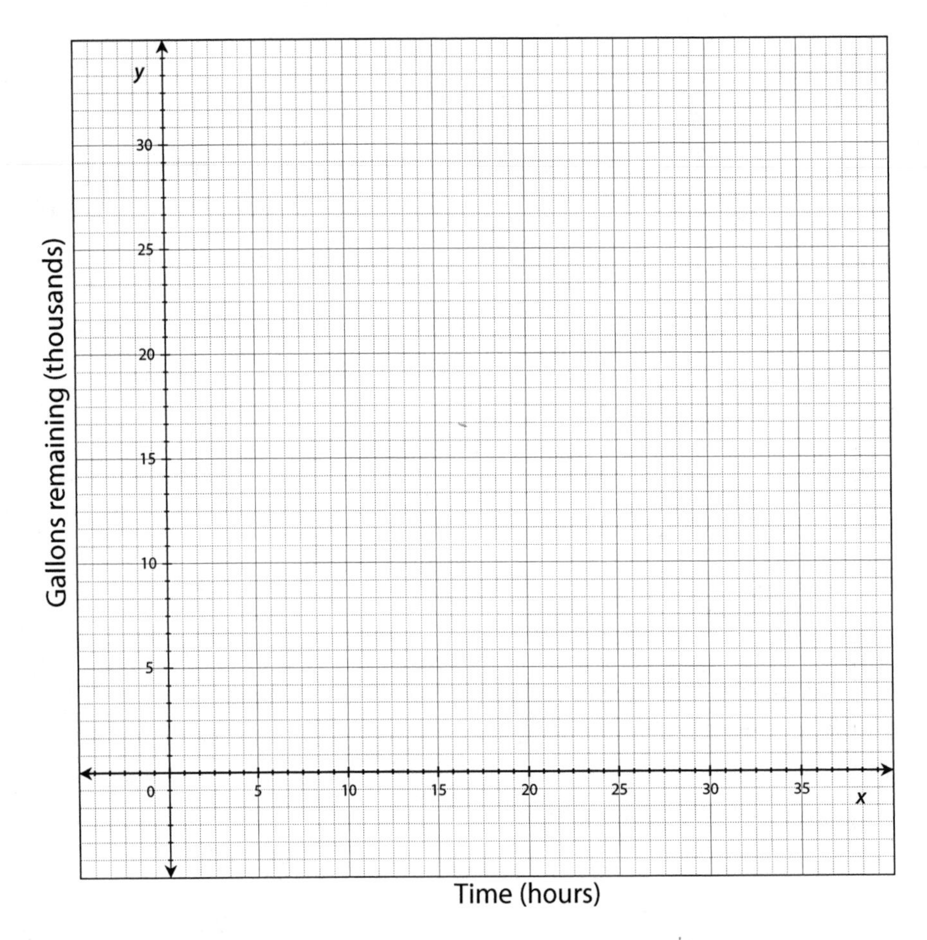

3. Does this situation represent a function? Explain your reasoning.

NAME:

FUNCTIONS
CCSS 8.F.2

Cars and Drivers

The U.S. Department of Transportation records the number of licensed drivers as well as the number of registered vehicles each year. Both sets of data can be modeled by linear functions. The function $y = 2.8x - 5401$ represents the number of licensed drivers in the United States between the years 1960 and 2010. The graph below represents the number of registered vehicles from 1960 to 2010.

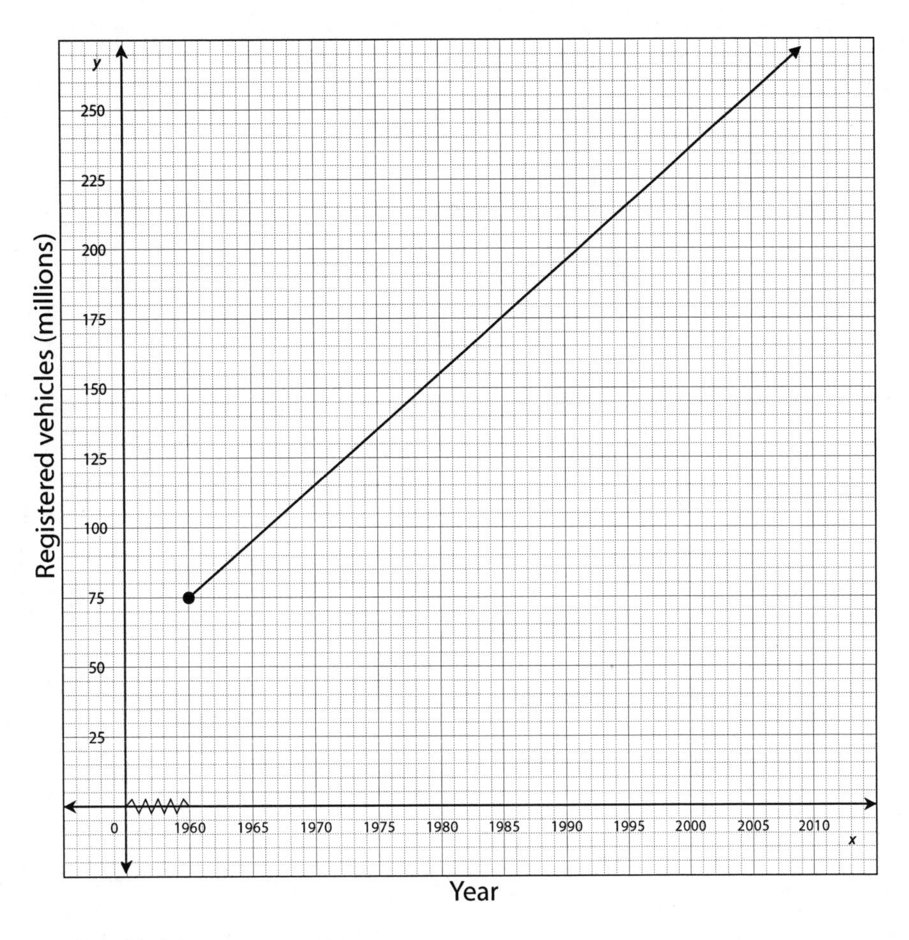

1. Which model has a greater rate of change?

2. According to the models, were there more licensed drivers or more registered vehicles in 1965?

3. If the number of licensed drivers and registered vehicles were to continue to follow these models, would there be more licensed drivers or more registered vehicles in 2020? Explain your answer.

NAME:

FUNCTIONS
CCSS 8.F.3

Slope-Intercept Form

Functions are usually written with y to the left of the equal sign. This form of an equation is called the slope-intercept form. It is written $y = mx + b$, where m is the slope of the line and b is the value at which the line crosses (intercepts) the y-axis. When a line crosses the y-axis, the x value is always zero.

Put each equation below into slope-intercept form.

1. $5x + y = 14$

2. $2x - y = 32$

3. $40 = 3x + y$

4. $7x - y = 9$

5. $5y = 5x - 25$

6. $y + 3x = 50$

7. $4y + 2x = 100$

8. $\frac{1}{3}y - \frac{1}{4}x = 5$

Slope-Intercept Equations

Simplify the following equations so that you could enter them more easily into a graphing calculator. Give the slope and *y*-intercept for each without graphing it.

1. $y - 3(x - 7) = 5x$

2. $y = -2 + (x + 1)$

3. $y + 4 = \frac{5}{3}(x + 6)$

4. $y = 28 + 2.5(x - 6)$

5. $y = 13.2(x - 20) + 125.6$

NAME:

FUNCTIONS
CCSS 8.F.4

How Long? How Far?

The Carmona family lives in Minnesota. They are driving to Florida for a vacation at an average speed of 60 miles per hour. Write an equation for a rule that can be used to calculate the distance they have traveled after any given number of hours. Then write a brief letter to the Carmona family that describes the advantages of having an equation, a table, and a graph to represent their situation.

Game Time

Abby and her brother Harry like to play a game called "U-Say, I-Say." Harry gives the "U-Say" number (an integer between –10 and +10). Abby has a secret rule she performs on the number that results in the "I-Say" number. Complete the table below by giving the missing "I-Say" values. Then describe Abby's rule in words and symbols.

U-Say	3	0	–4	1	2	5
I-Say	11	2	–10	5		

Hedwig's Hexagons

Hedwig is using toothpicks to build hexagon patterns. The first three are pictured. How do you think her pattern will continue? She has begun to collect information in a table in order to explore the relationships among the number of toothpicks, the number of hexagons, and the length of the outside perimeter of the figure. Use the diagram to complete each problem that follows.

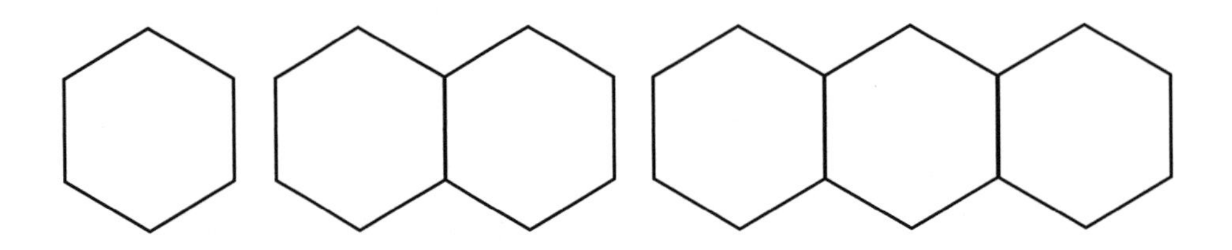

1. Complete Hedwig's table.

Number of toothpicks	6		
Number of hexagons	1	2	
Perimeter of figure	6	10	

2. Now write an equation for the relationship between the number of hexagons in each figure and the perimeter of the figure.

3. Write another equation for the relationship between the number of hexagons and the number of toothpicks.

NAME:

FUNCTIONS
CCSS 8.F.4

Finding Slope-Intercept Equations

Find an equation written in slope-intercept form that satisfies each condition below.

1. a line whose slope is -3 and passes through the point (2, 5)

2. a line whose slope is $-\frac{2}{3}$ and passes through the point (4, -1)

3. a line that passes through the points (2, 6) and (6, 1)

4. a line that passes through the points (-2, 3) and (4, -3)

5. a line that passes through the points (-1, -1) and (-5, -5)

NAME:

FUNCTIONS

CCSS 8.F.4

Up and Down the Line I

Write an equation for the line pictured in each graph below.

1.

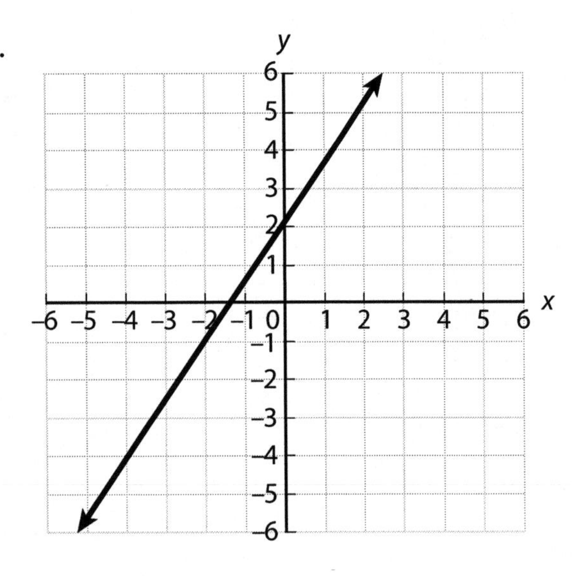

2.

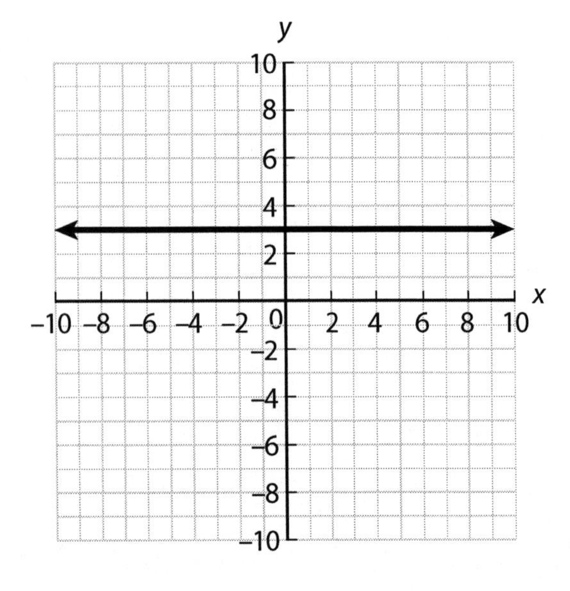

3.

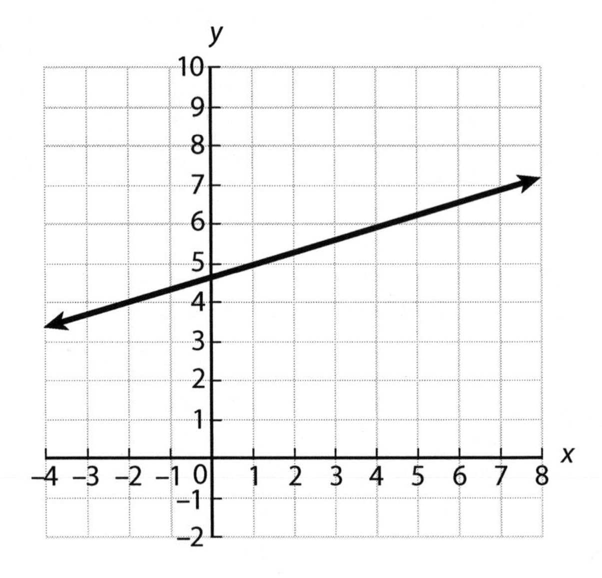

4.

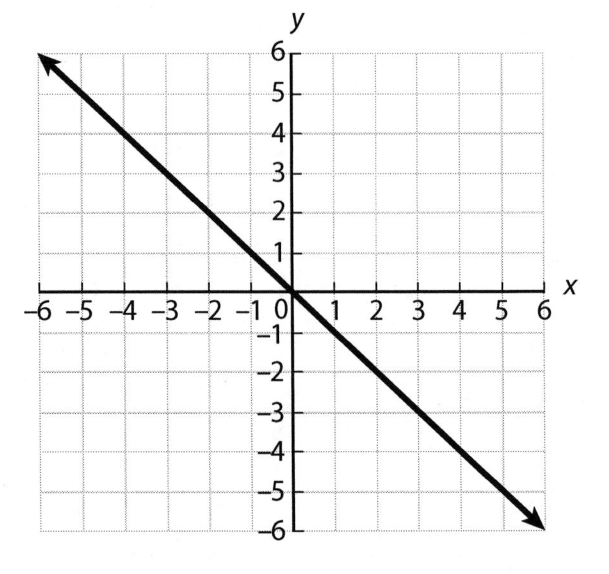

Up and Down the Line II

Write a linear equation for each condition below.

1.

x	–1	0	1	2	3
y	1	3	5	7	9

2. a line whose slope is –3 and y-intercept is 5

3. a line that passes through the points (2, 5) and (5, 6)

4. a line that passes through the point (3, 7) and has a slope of $\frac{2}{3}$

Butler Bake Sale

The Butler Middle School eighth grade class is planning a bake sale as a fall fund-raiser. Luis is chairing the planning committee. He gave a brief survey to determine what price should be charged for each brownie. He predicts from his results that a price of $0.50 per brownie will result in 200 brownies sold. He also predicts that a price of $1.00 per brownie will result in about 50 brownies sold. He is assuming that the relationship between the brownie price and the number sold is a linear relationship.

1. Write an equation for the relationship that Luis has predicted between the cost and the number of brownies sold.

2. Find the slope and y-intercept. Then explain what these mean in the context of Luis's information.

3. What if the committee decides to charge $0.70 per brownie? How many can they expect to sell?

4. Luis has taken an inventory and found that the class has 300 brownies to sell. Using your equation, what would be an appropriate price to charge for each brownie?

NAME:

FUNCTIONS
CCSS 8.F.4

Looking for Lines

Explain how you can find the equation of a line if you know the information below. Use examples to explain your thinking.

1. the slope and y-intercept

2. two points on the line

3. the slope of the line and a point that is on the line, but is not the y-intercept

NAME:

FUNCTIONS
CCSS 8.F.4

Thinking About Equations of Linear Models

Think about ways to find an equation of the form $y = mx + b$ or $y = a + bx$ from a table of data or a graph of the points. How can you find the equation if you know the slope and y-intercept? How can you find the graph by looking for the rate of change and other values from the table? How can you find the equation of the line if the slope and y-intercept are not given? Write a few sentences to explain your thinking.

NAME:

FUNCTIONS
CCSS 8.F.4

Thinking About Change and Intercepts in Linear Models

Think about linear models in various forms. How can you see or find the rate of change in a linear model from a table of values? How would you find it from the graph? How does it appear in the equation? How would you determine the *y*-intercept in those three situations? Write a few sentences to explain your thinking.

NAME:

FUNCTIONS
CCSS 8.F.4

Freezing or Boiling?

Lucas knows that the relationship between Celsius and Fahrenheit temperatures is a linear one. However, he often forgets the equation used to convert Celsius temperatures to Fahrenheit. He remembers that when water freezes, the temperatures are (0°C, 32°F). When water boils, the respective temperatures are (100°C, 212°F). Explain to Lucas how he can find the equation knowing that the relationship is a linear one.

Border Tiles for Goldfish Pools

Landscapers often use square tiles as borders for garden plots and pools. The drawings below represent square pools for goldfish surrounded by 1-foot square tiles. For example, if the square pool is 2×2, there are 12 tiles in the border. If the square pool is 3×3, there are 16 square tiles in the border.

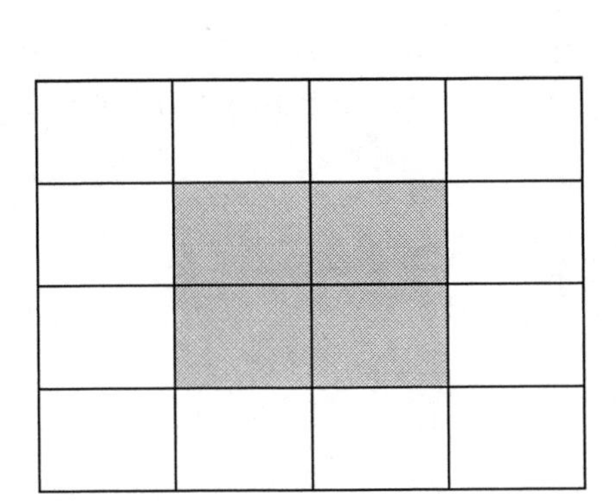

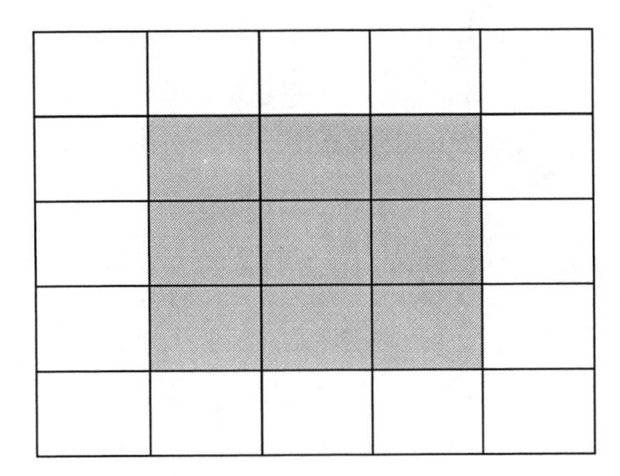

Collect data for a number of different-sized square pools. How many square tiles would be in the border around a pool that is 5×5? A pool that is 10×10? What patterns do you see in your data? Show and explain your thinking. Graph the data that you have collected. Then describe your graph.

NAME:

FUNCTIONS
CCSS 8.F.4

Table Patterns

Look at the tables below. For each table, do the following:

a. Describe symbolically or in words the pattern hiding in the table.

b. Give the missing values in the table.

c. Tell whether the relationship between x and y in the table is quadratic, linear, inverse, or exponential, or if it's some other type of relationship. Explain how you know.

1.

x	–1	0	1	2	3	4	5
y	5	7	9	11	13		

2.

x	0	1	2	3	4	5
y	1	2	5	10		

Parachuting Down

Kamilla took her first parachute jump lesson last weekend. Her instructor gave her the graph below that shows her change in altitude in meters during a 2-second interval. Use the graph to answer the questions that follow.

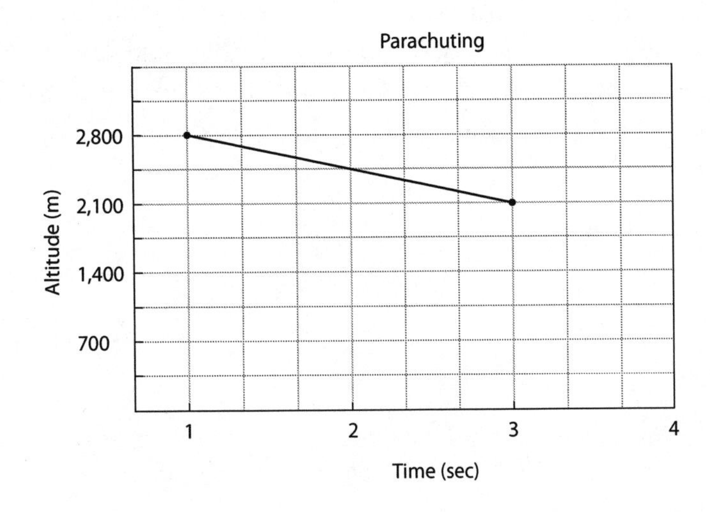

1. What is the slope of the line segment?

2. Estimate Kamilla's average rate of change in altitude in meters per second.

3. Give the domain and range for this graph.

Sketching a Graph

Sketch a graph that represents each situation below. Identify your variables and label the axes of your graph accordingly.

1. Edgar is paddling his kayak down the reservoir to visit a friend. For half an hour, he paddled slowly, enjoying the morning. Then he noticed that the weather might be changing and increased his speed. After another 10 minutes, the wind picked up. For about 20 minutes, the wind blew in the direction he was paddling. Then it shifted and blew directly into his face for the last 15 minutes.

2. An ice cream shop keeps track of ice cream sales using a graph. Sales are higher during warm weather and lower during cool weather. The shop opens its doors every year on the first day of spring. This year, the first few days of spring were very cold and rainy. Then, the weather slowly warmed each day and the days stayed dry. After three weeks, the weather turned colder again and it was cloudy for a week and a half.

NAME:

FUNCTIONS
CCSS 8.F.5

Interpreting Graphs I

Which graph best represents each given situation? Be prepared to justify your answers.

1. A commuter train pulls into a station and drops off all its passengers.

a.

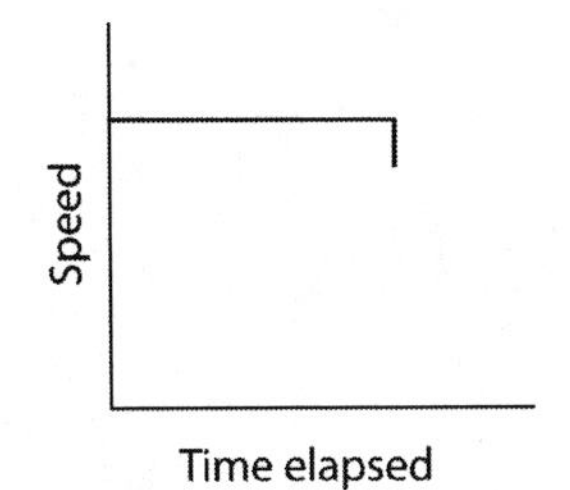

c.

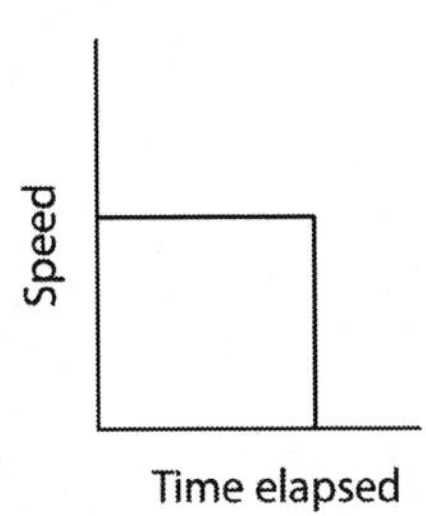

b.

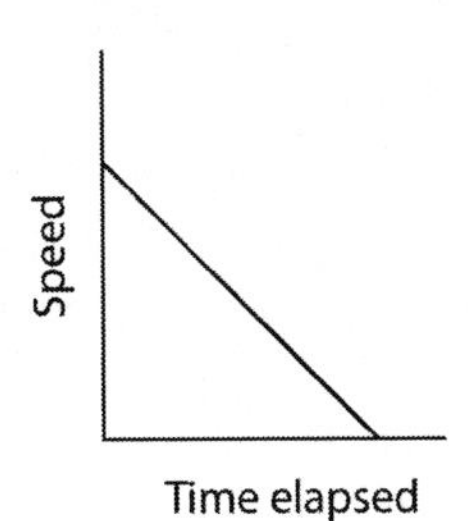

d.

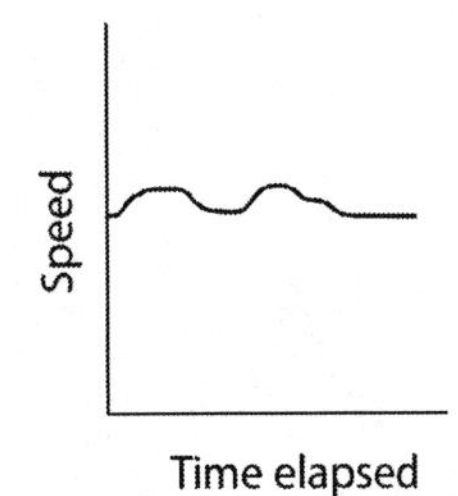

2. A child walks to a slide, climbs up to the top, and then slides down.

a.

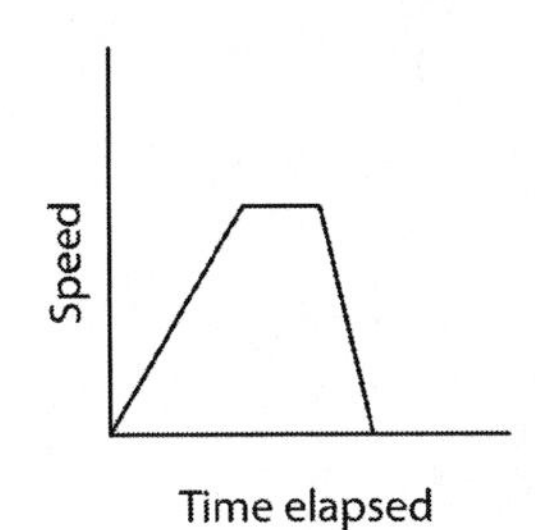

c.

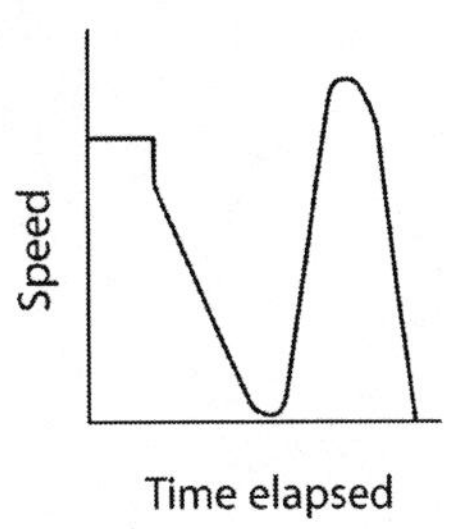

b.

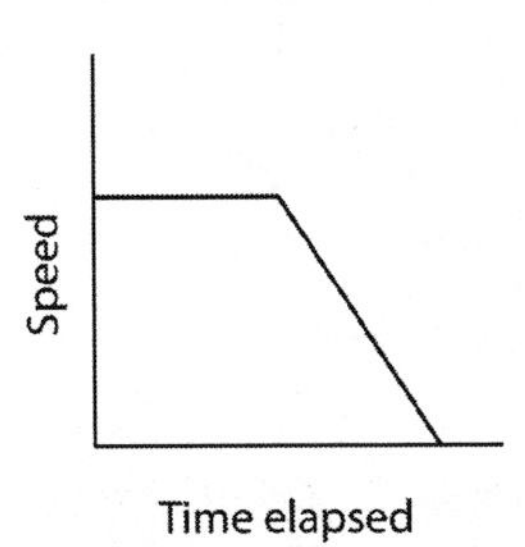

d.

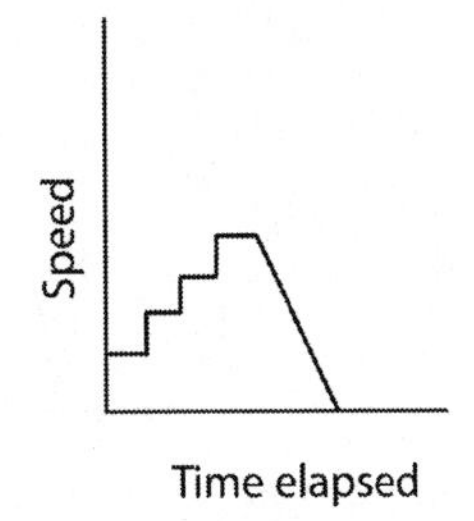

Interpreting Graphs II

Which graph best represents each given situation? Be prepared to justify your answers.

1. A little boy swings on a swing at a playground.

a.

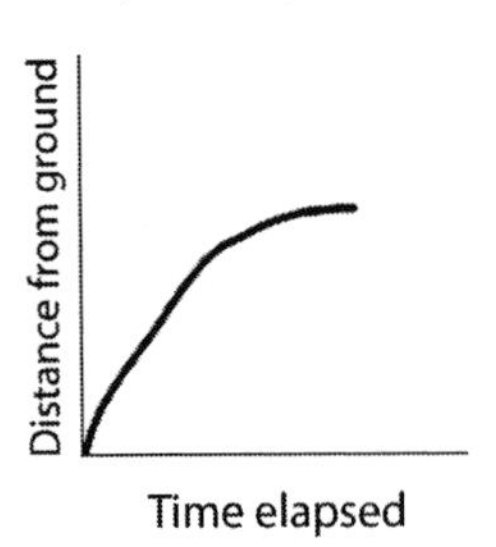

c.

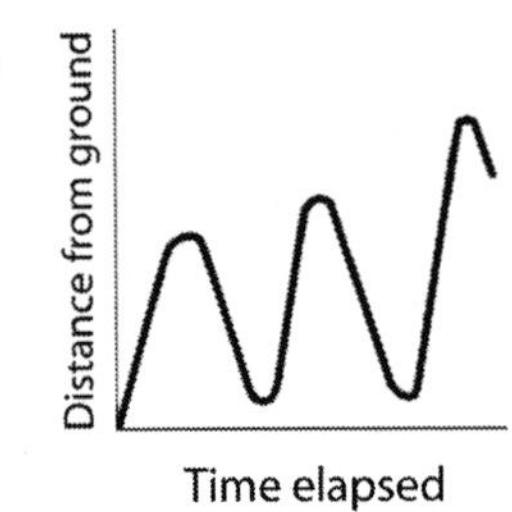

b.

d.

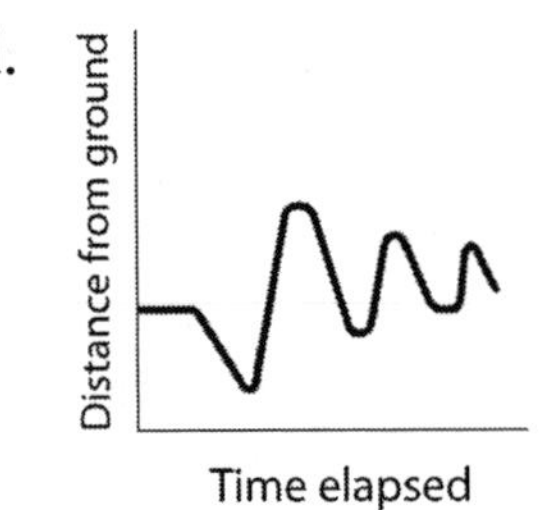

2. A woman walks up a hill at a constant rate and then runs down the other side.

a.

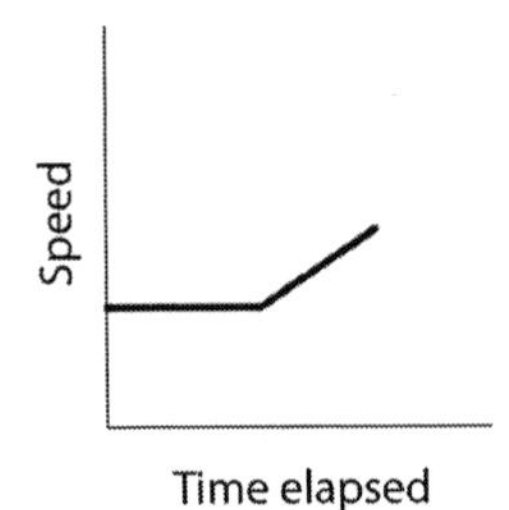

c.

b.

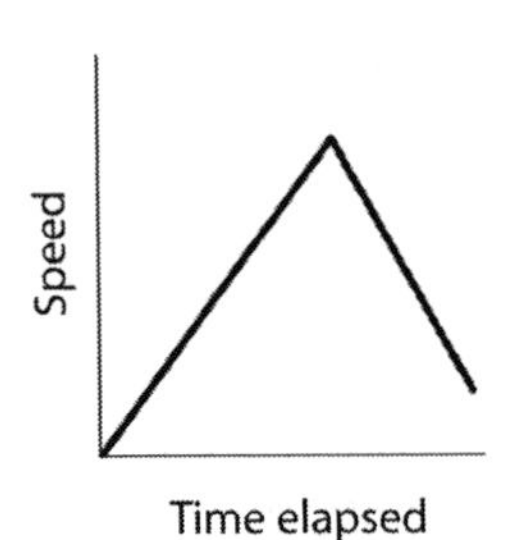

d.

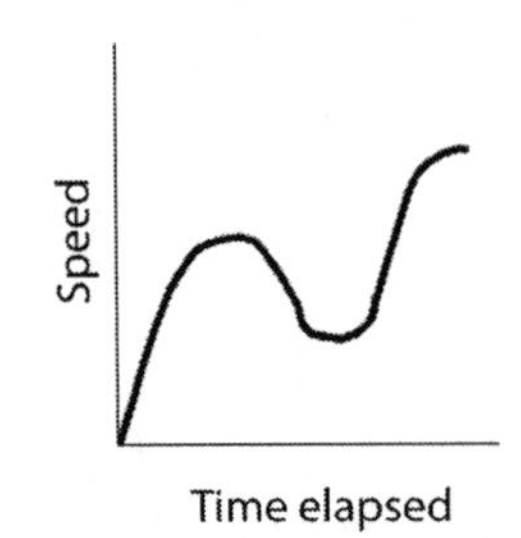

NAME:

FUNCTIONS
CCSS 8.F.5

Ferris Wheel Graphs

Enrique took a ride on a Ferris wheel at the amusement park last weekend. He has sketched these graphs to represent that ride. Which one best represents the idea that he's trying to show? Explain why you chose that graph.

a.

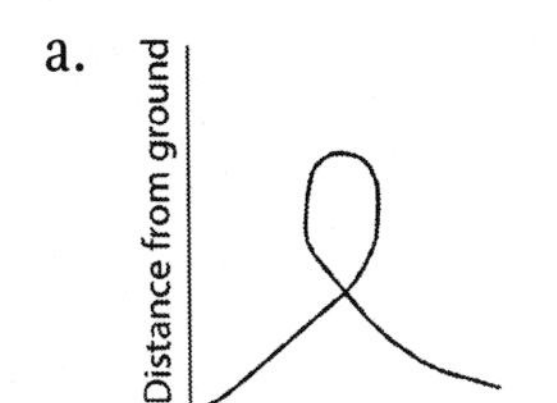

b.

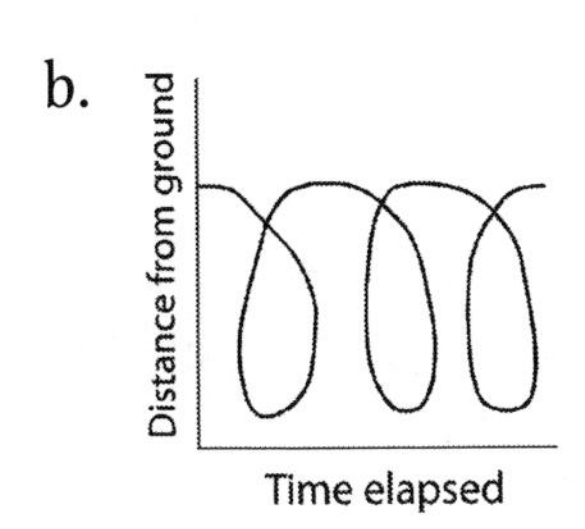

c.

Distance from ground

Time elapsed

d.

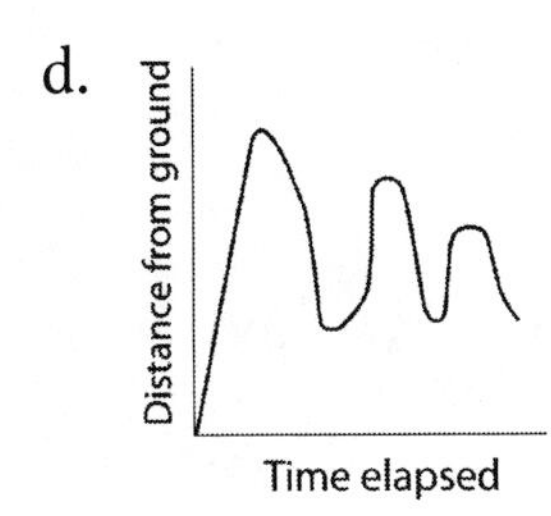

NAME:

FUNCTIONS
CCSS 8.F.5

Roller Coaster Ride

Marjorie and Jack love to ride roller coasters and then portray their ride in a graph. Last Saturday, they rode the Mighty Twister and sketched the following graph of that experience.

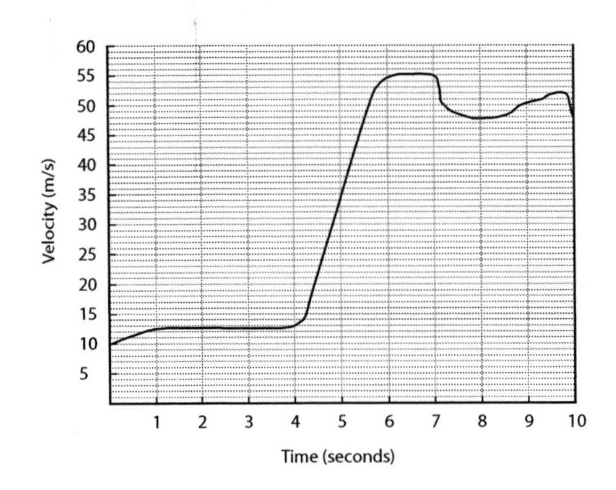

1. What does the pattern of their graph tell you about when they were going down the steepest hill?

2. What was their approximate velocity 1 second into the ride? What was their velocity 8 seconds into the ride?

3. After getting started, when were they going the slowest? The fastest?

4. Describe what was happening to them between 4 and 8 seconds.

NAME:

FUNCTIONS
CCSS 8.F.5

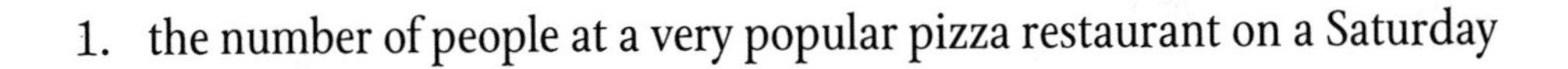

Graphing People Over Time

Think about each situation below. Then sketch a graph to represent the situation over a 24-hour period of time. Label each graph carefully using the horizontal axis to represent the time of day. Be prepared to explain and justify your choices.

1. the number of people at a very popular pizza restaurant on a Saturday

2. the number of people in a school building on a weekday in September

3. the number of people at a sports stadium on the day of a big game

4. the number of people at a movie theater on a weekend day

Renting Canoes

Dan and Julia have reservations for 10 students for the Outdoor Adventure Club's annual trip. This year they are planning a canoe camping trip. The Otter Mountain River Livery rents canoe and camping gear for $19 per person. Dan and Julia expect no more than 50 students in all to go on the trip. Using increments of 10 campers, make a table showing the total rental charge for 10 to 50 campers. Then make a coordinate graph of the data.

NAME:

FUNCTIONS
CCSS 8.F.5

Grow Baby!

The average growth weight in pounds of a baby born in the United States is a function of the baby's age in months. Look at the sample data provided and graph the information in an appropriate window. Then write two sentences about the graph and the relationship between the two variables.

Age in months	Weight in pounds
0	7
3	13
6	17
9	20
12	22
15	24
18	25
21	26
24	27

NAME:

GEOMETRY
CCSS 8.G.2

Double Trouble

Look at the two triangles below. Are the two triangles similar, congruent, or both? Discuss your ideas with a partner, and write your answer below.

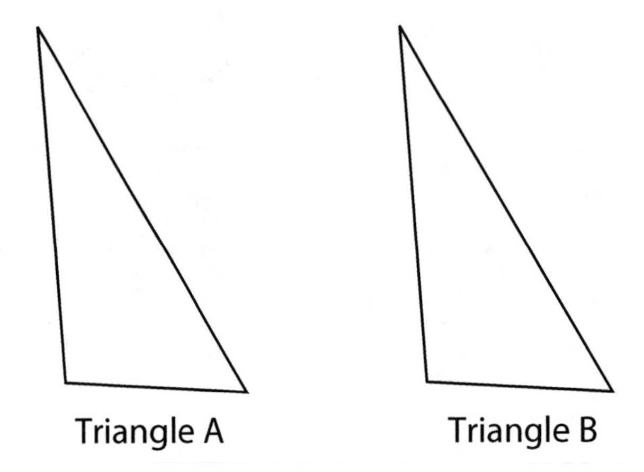

NAME:

GEOMETRY
CCSS 8.G.3

Up, Down, and All Around

There are five triangles in the diagram below. Triangle 1 is the original triangle. The other four triangles came from changing Triangle 1 in some way.

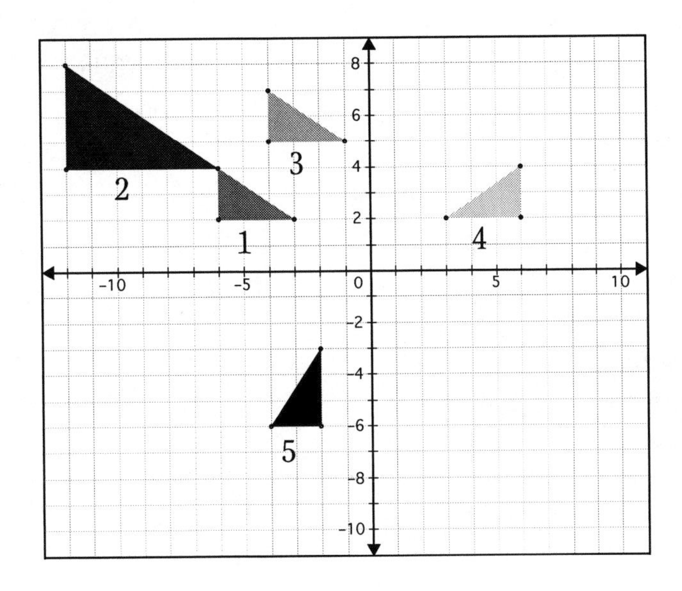

Work with a partner to finish the following statements about the triangles in the diagram above.

1. To change Triangle 1 into Triangle 4, ______________________________

   ____

2. To change Triangle 1 into Triangle 3, ______________________________

   ____

3. To change Triangle 1 into Triangle 5, ______________________________

   ____

4. To change Triangle 1 into Triangle 2, ______________________________

   ____

NAME:

GEOMETRY
CCSS 8.G.3

Translate This!

Answer the questions below, sketching the figures if needed.

1. If you drew a rectangle on a sheet of paper and then translated it perpendicular to the paper (moving it up in space), what three-dimensional shape would you make?

2. If you picked up a circular coaster and rotated it around its center 360 degrees, what three-dimensional shape would you make?

3. If you drew a triangle on a sheet of paper and then translated it perpendicular to the paper (moving it up in space), what three-dimensional shape would you make?

4. If you put a cylinder on one of its bases on a table and then imagined flattening it out from above, what two-dimensional shape would you get?

NAME:

GEOMETRY
CCSS 8.G.4

Mirror, Mirror on the Wall

In the diagram below, triangles *ABC* and *XYZ* are mirror images of each other reflected over the *y*-axis.

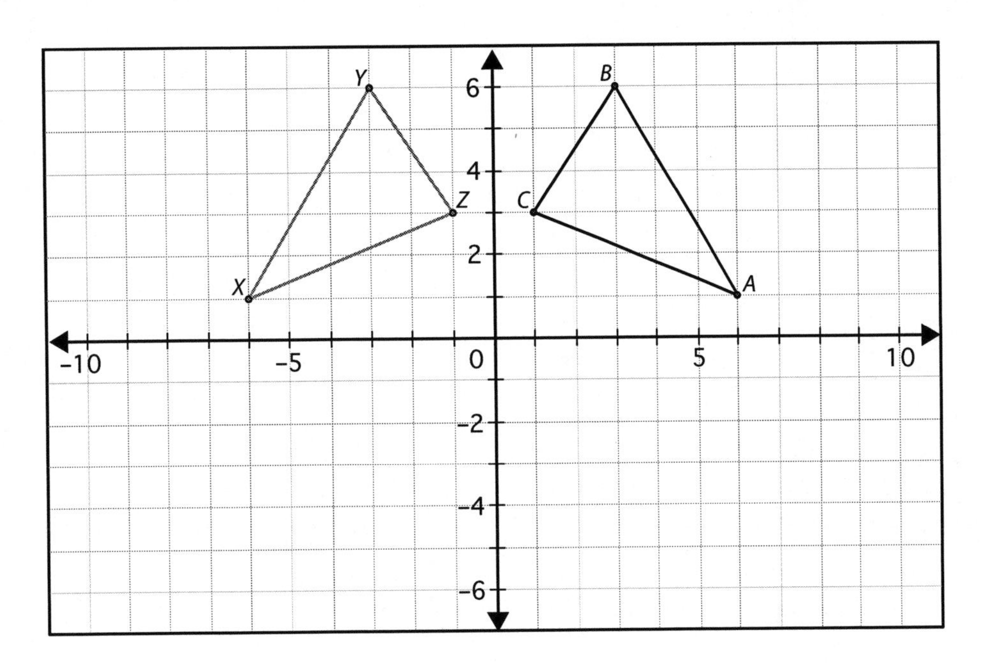

1. What are the coordinates of point *X*?

2. What are the coordinates of point *A*?

3. What are the coordinates of point *Y*?

4. What are the coordinates of point *B*?

5. What are the coordinates of point *Z*?

6. What are the coordinates of point *C*?

7. What do you notice about the coordinates of the points that are reflections of each other? Refer to the *x*-coordinate and the *y*-coordinate.

NAME:

GEOMETRY
CCSS 8.G.4

Copy Cat

Examine the two triangles below. Triangle *ABC* is similar to triangle *XYZ* and is $\frac{1}{2}$ the size.

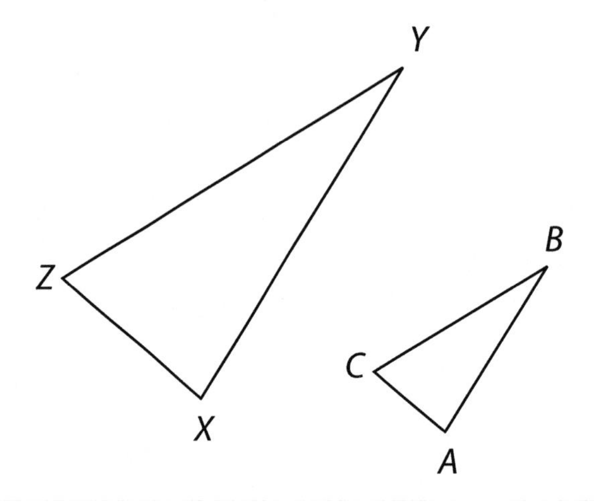

1. Which side of triangle *ABC* is in the same place as side *XY*?

2. Which side of triangle *ABC* is in the same place as side *YZ*?

3. Which side of triangle *ABC* is in the same place as side *ZX*?

4. Do angles *A* and *X* appear to be the same size?

5. Do angles *B* and *Y* appear to be the same size?

6. Do angles *C* and *Z* appear to be the same size?

7. How do the areas of the two triangles compare to one another?

NAME:

GEOMETRY
CCSS 8.G.4

Tree Dilemma

Eric's neighbor wants to cut down a dead tree that is in his yard. Eric is worried that when the tree is cut, it will fall on his garage, which is 42 feet from the tree. His neighbor decides to measure the height of the tree by using its shadow. The tree's shadow measures 47.25 feet. At the same time, Eric puts a yardstick next to the tree, and the yardstick casts a shadow of 3.5 feet. Will the tree hit Eric's garage if it falls the wrong way? Explain carefully. Include a sketch of the situation to help clarify your thinking.

NAME:

GEOMETRY
CCSS 8.G.7

Road Trip

Use what you know to complete each problem that follows.

1. Two cars leave the same parking lot in Sandy Springs at noon. One travels due north, and the other travels directly east. Suppose the northbound car is traveling at 60 mph and the eastbound car is traveling at 50 mph. Make a table that shows the distance each car has traveled and the distance between the two cars after 1 hour, 2 hours, 3 hours, and so forth. Describe how the distances are changing.

2. Suppose the northbound car is traveling at 40 mph, and after 2 hours the two cars are 100 miles apart. How fast is the other car going?

3. Draw a diagram to help explain the situation. Explain your thinking clearly.

NAME:

GEOMETRY
CCSS 8.G.9

Comparing Cylinders

Two cylinders are pictured below. All the dimensions of cylinder A are 3 times the dimensions of cylinder B.

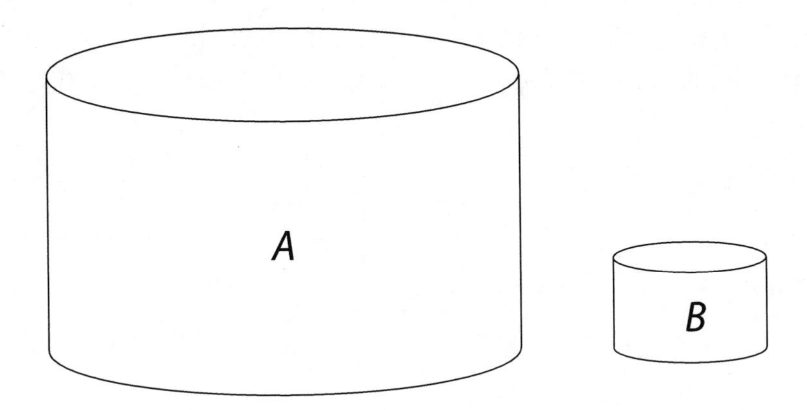

1. What is the ratio of the radius of cylinder A to the radius of cylinder B?

2. What is the ratio of the height of cylinder A to the height of cylinder B?

3. What is the ratio of the surface area of cylinder A to the surface area of cylinder B?

4. What is the ratio of the volume of cylinder A to the volume of cylinder B?

Concession Concern

Emile and Sharonda are in charge of concessions for their school's football games. They usually sell popcorn in cylindrical containers like the one below on the right. However, they have just noticed that their supply of these containers is gone, and there's no time to get more before the next game. Emile notices that there are plenty of the cone-shaped containers like the one on the left. He suggests that they use those. Both of the containers have a height of 8 inches and a radius of 3 inches.

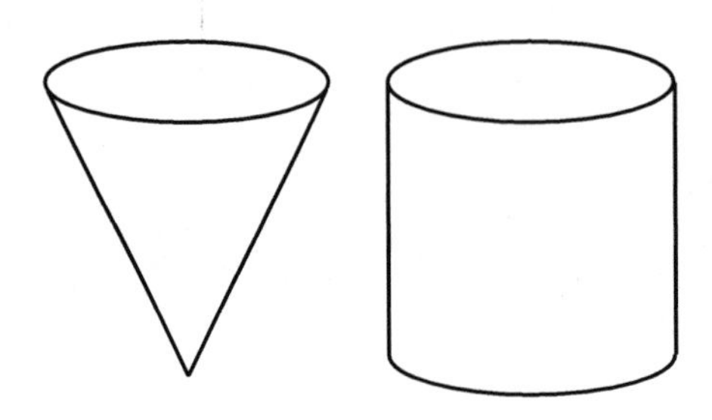

1. What is the difference between what the cone will hold and what the cylinder will hold?

2. If Emile and Sharonda usually charge $1.50 for the cylinder full of popcorn, what should they charge for the cone?

NAME:

GEOMETRY
CCSS 8.G.9

Popcorn Pricing

Sisters Elizabeth and Emma enjoy going to the movies on Saturday afternoons. Sometimes they each buy a small popcorn for $3.00, and sometimes they buy a large popcorn that costs $6.00 and share it. Both containers are cylinders. The heights of the two containers are the same. However, the radius of the large container is about twice the radius of the small container. Which purchase gives the sisters the most popcorn for their money? Draw a sketch of the two popcorn containers, and explain your thinking.

NAME:

GEOMETRY
CCSS 8.G.9

Juice Packaging

The Johnny Appleseed Juice Company is changing its packaging. The juice currently comes in a cylindrical container like the one pictured below on the right. The company is considering changing this container to a prism, like the one pictured below on the left. The prism will have a square base that has the same width as the diameter of the cylinder, which is 5 inches. The height of each container is $7\frac{1}{2}$ inches.

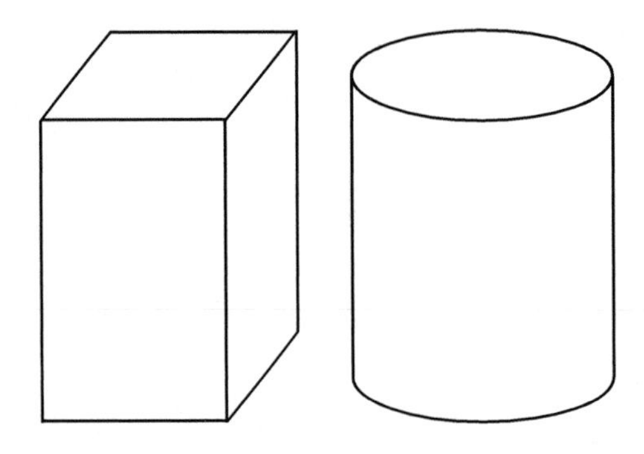

1. What is the volume of each container?

2. If the company was charging $3.29 for the cylindrical container of juice, what would be a fair price to charge for the new container?

Cylinder Expressions

The cylinder below has a base area of 45 cm². It is partially filled with liquid up to a height of x cm. The height from the top of the liquid to the top of the cylinder is represented by y cm. The total height is represented by h cm. Which expressions below properly represent the total volume of the cylinder? Briefly explain your reasoning.

a. $45(h - x)$

b. $45(x + y)$

c. $45x + 45y$

d. $45xy$

NAME:

GEOMETRY
CCSS 8.G.9

Paper Cylinders

Erica has a sheet of paper that is $8\frac{1}{2}$ inches by 11 inches. Without cutting the paper, she wants to make a container with the greatest possible volume. (She will make the top and bottom of the container with another sheet of paper.) She thought of rolling the paper to make an open-ended cylinder and realized that there are two ways to do this. Her friend Aleah suggests folding the paper to make a rectangular prism with square ends instead. Erica points out that there are also two ways to fold the paper to make the sides of a prism with a square base.

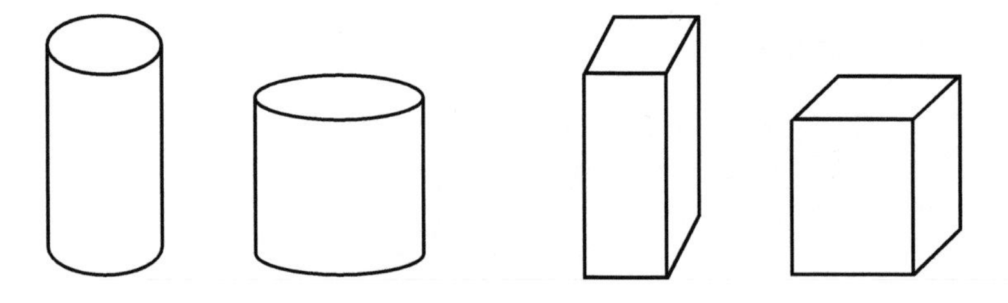

1. Predict which of the four containers has the greatest volume. You may want to make models of the containers to help explain your reasoning.

2. Now find the volume of each container. What is the volume of the largest container?

3. How much greater is the volume of that container than the volume of the other container of the same height?

4. Write a note to Erica that explains and justifies your thinking clearly.

NAME:

GEOMETRY
CCSS 8.G.9

Tennis Ball Packaging

Rochelle works as a packaging engineer for the Creative Carton Company. The company wants to remove the air from containers of tennis balls so the balls will retain a good bounce. Rochelle needs to know how much air there is in a standard container of tennis balls. Help Rochelle by finding the amount of empty space in a cylindrical container that is 18 centimeters tall and contains 3 tennis balls that are each 6 centimeters in diameter. Show your work and explain how you found your answer.

NAME:

STATISTICS AND PROBABILITY
CCSS 8.SP.1

Analyzing Scatter Plots

Ashley is reviewing the results of data collected and graphed from three different experiments. Help her by describing the patterns you see in these graphs. If appropriate, draw lines that you think might fit the data. If you don't think that a line fits the data, explain why.

Graph A

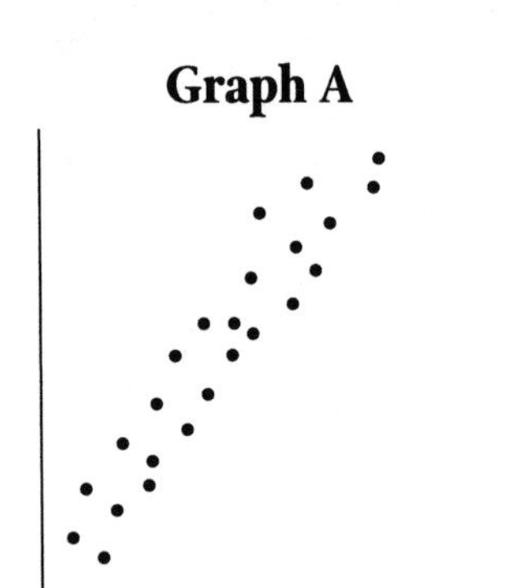

Graph B

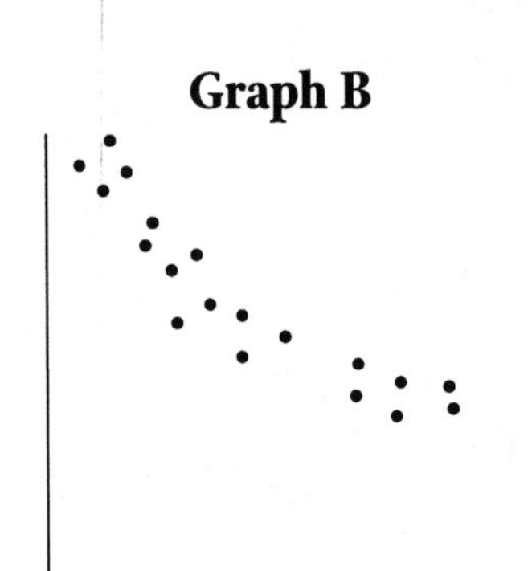

Graph C

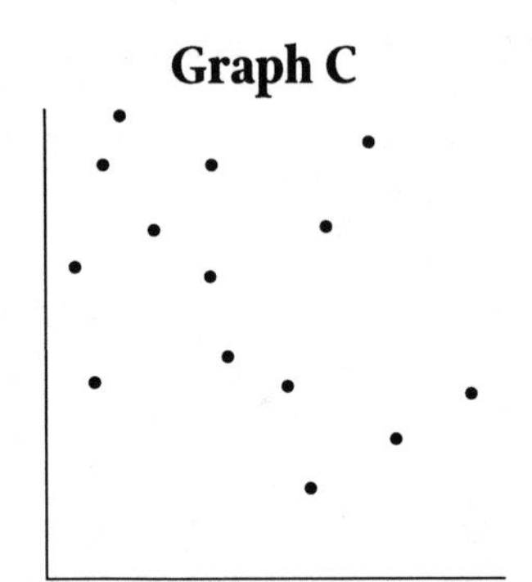

NAME:

STATISTICS AND PROBABILITY

CCSS 8.SP.2

Mr. Wiley's Baby

Mr. Wiley presented some data to his algebra class about the early growth of his new baby boy. Hayley created a scatter plot for Mr. Wiley's data. The data and scatter plot are shown below.

Week	Weight
1	8.5
2	9.25
3	9.75
4	9.75
5	10.5
6	11
7	11.5
8	11.75

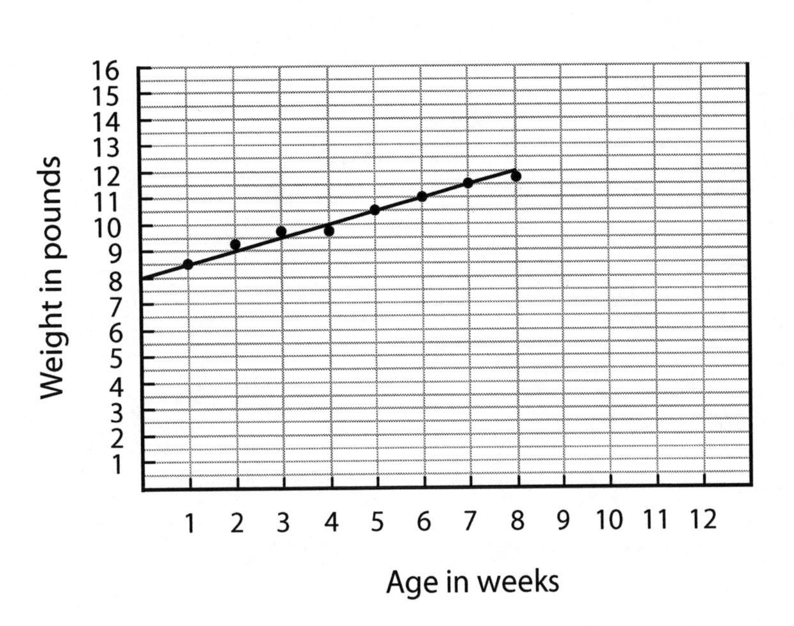

1. Does the data seem to represent a linear model? What equation fits Hayley's graph model?

2. How long do you suppose this growth pattern will continue in this way? If it does continue, what would you expect Mr. Wiley's son to weigh in 1 year? In 2 years? In 14 years?

3. Within what age limits do you think this is a reasonable representation of the growth of Mr. Wiley's baby?

Answer Key

The Number System

Irrational Numbers, p. 1

Yes, it's irrational. It's a decimal that continues without repeating.

Where Do They Go?, p. 2

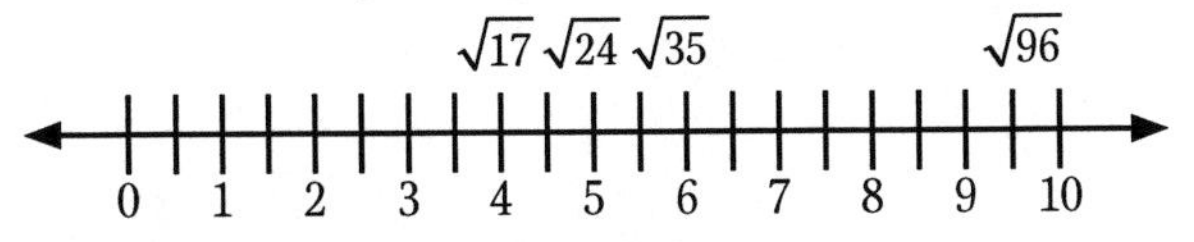

Expressions and Equations

Tearing and Stacking Paper, p. 3

$2^{50} = 1.125899907 \times 10^{15}$ sheets of paper /1,000 = $1.125899907 \times 10^{12}$ inches /12 = $9.382499224 \times 10^{10}$ feet /5,280 = 17,769,884.89 miles

Linear vs. Nonlinear, p. 4

1. nonlinear
2. linear
3. nonlinear
4. nonlinear
5. linear
6. nonlinear

Graphing Linear Functions, p. 5

1. graph of ordered pairs (0, –2), (1, 0), (2, 2), (3, 4), (4, 6), (5, 8)
2. graph of ordered pairs (0, 1), (1, 2), (2, 3), (3, 4), (4, 5), (5, 6)

The Seesaw Problem, p. 6

1. Caleb weighs about 84 pounds.
2. If Alex and Samuel continue to sit at 3 feet, then $140 \times 3 = 84 \times D$, and $D = 5$ feet. If the seesaw is long enough, then Caleb can balance them. Otherwise, he cannot.

Systems of Linear Equations, p. 7

Answers will vary.

1. Sample answer: to find an ordered pair that satisfies both equations
2. Sample answer: Graph both lines on the same set of axes, and locate the intersection point of the two lines.
3. Sample answer: Substitute values of x into both equations. Look for identical y values for the same value of x.
4. Sample answer: solve algebraically by substitution or elimination
5. Sample answer: If the values in the table have the same constant rate of change, if the lines are parallel, or if the slopes are the same but have different y-intercepts, then there is no solution.

Exploring Systems of Equations, p. 8

1. a. Check graphs for accuracy.

   b.

x	0	1	2	3	4	5
y	–3	–1	1	3	5	7

2. a. Check graphs for accuracy.

   b.

x	0	1	2	3	4	5
y	12	9	6	3	0	–3

3. (3, 3)
4. (3, 3)
5. The point satisfies both equations.
6. The point only works in the first equation.
7. Any other point chosen will only work in one of the equations, not in both equations.

Nathan's Number Puzzles I, p. 9

Word puzzles will vary.

1. (2, 1)
2. (–2, 4)
3. not possible

Understanding Systems of Linear Equations, p. 10

Students' choices and rationales may vary.

1. (–4, –5)
2. (–6, 4)
3. (–20, 10)
4. (3.6, 2.4)

Nathan's Number Puzzles II, p. 11

1. $x + y = 10$; $x + 2y = 8$; (12, –2)
2. $x + 2y$; $x + y = 15$; (10, 5)
3. $x - y = 2$; $2x - 2y = 4$; Any values for x and y that differ by 2 will work in this puzzle. There are infinite solutions.

Functions

Contaminated Drinking Water, p. 12

1. 24,975; 20,350; 15,725

2.

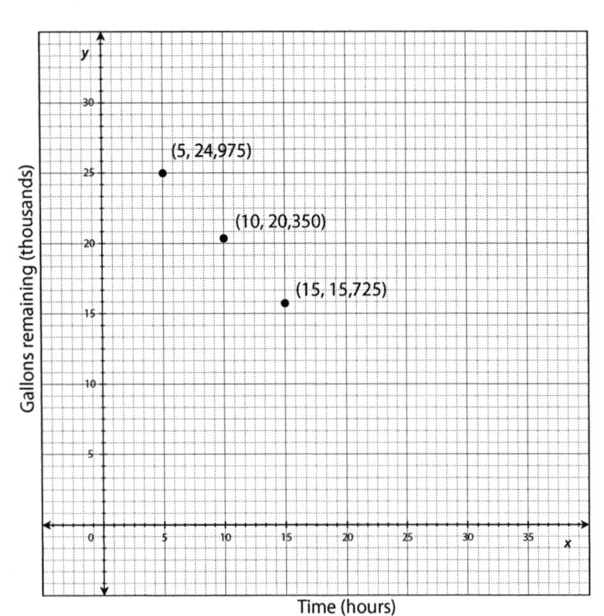

3. This situation represents a function because for each input, there is exactly one unique output.

Cars and Drivers, p. 13

1. The rate of change for the function representing the number of licensed drivers is 2.8. The rate of change for the function representing the number of registered vehicles is 4. The model representing registered vehicles has the greater rate of change.
2. In 1965, there were more licensed drivers than registered vehicles.
3. If the number of licensed drivers and registered vehicles continued to follow these models, there would be more registered vehicles than licensed drivers in 2020.

Slope-Intercept Form, p. 14

1. $y = -5x + 14$
2. $y = 2x - 32$
3. $y = -3x + 40$
4. $y = 7x - 9$
5. $y = x - 5$
6. $y = -3x + 50$
7. $y = -1/2\,x + 25$
8. $y = 3/4\,x + 15$

Slope-Intercept Equations, p. 15

1. $y = 8x - 21$; slope = 8, y-intercept = (0, 21)
2. $y = x - 1$; slope = 1, y-intercept = (0, –1)
3. $y = \frac{5}{3}x + 6$; slope = 5/3, y-intercept = (0, 6)
4. $y = 2.5x + 13$; slope = 5/2, y-intercept = (0, 13)
5. $y = 13.2x - 138.4$; slope = 66/5, y-intercept = (0, –138.4)

How Long? How Far?, p. 16

$D = 60t$. Letters will vary.

Game Time, p. 17

U-Say	3	0	–4	1	2	5
I-Say	11	2	–10	5	8	17

Abby's rule: I-Say = 3 • U-Say + 2

Hedwig's Hexagons, p. 18

1.

Number of toothpicks	6	11	16
Number of hexagons	1	2	3
Perimeter of figure	6	10	14

2. $T = 5H + 1$
3. $P = 4H + 2$

Finding Slope-Intercept Equations, p. 19

1. $y = -3x + 11$
2. $y = -\frac{2}{3}x + \frac{5}{3}$
3. $y = -\frac{5}{4}x + \frac{17}{2}$
4. $y = -x + 1$
5. $y = x + 0$

Up and Down the Line I, p. 20

1. $y = \frac{3}{2}x + 2$
2. $y = 3$
3. The line passes through (1, 5) and (4, 6), thus $y = \frac{1}{3}x + \frac{14}{3}$.
4. $y = -x$

Up and Down the Line II, p. 21

1. $y = 2x + 3$
2. $y = -3x + 5$
3. $(y - 5) = \frac{1}{3}(x - 2)$
4. $(y - 7) = \frac{2}{3}(x - 3)$

Butler Bake Sale, p. 22

1. $y = -300x + 350$ for x = cost and y = number sold
2. The slope = –300, and the y-intercept = 350. For every \$1.00 increase in price, the number sold decreases by 300; in addition, they can give away 350 brownies if they don't charge anything.
3. 140 brownies
4. about \$0.17 for each brownie

Looking for Lines, p. 23

1–3. Answers will vary.

Thinking About Equations of Linear Models, p. 24

Answers will vary. Sample answer: If you know the slope and y-intercept, you can substitute the values into the equation $y = mx + b$ or $y = a + bx$. The slope can be determined from the graph by finding the change in y divided by the change in x. The y-intercept will be the point where $x = 0$ in the table and where the line crosses the y-axis on the graph.

Thinking About Change and Intercepts in Linear Models, p. 25

Answers will vary. Sample answer: The rate of change is the value found when the difference in two y values is divided by the difference in the corresponding two x values, either from the table or from the graph using any two points. In the equation, the rate of change is the value m or b when the equation is written in the form $y = mx + b$ or $y = a + bx$. The y-intercept is the value $(0, x)$.

Freezing or Boiling?, p. 26

Slope is $\frac{212-32}{100-0} = \frac{180}{100} = \frac{9}{5}$, thus $y = \frac{9}{5}x + 32$.

Note: y = Fahrenheit and x = Celsius.

Border Tiles for Goldfish Pools, p. 27

Side of pool	Number of border tiles
2	12
3	16
4	20
5	24
6	28
7	32

Tiles increase by 4: $B = 4S + 4$. The graph is a straight line.

Table Patterns, p. 28

1. a. $y = 2x + 7$; As x increases, y increases by a constant amount. First differences are constant.
   b. missing values: (4, 15), (5, 17)
   c. The relationship is linear.
2. a. $y = -x^2 + 11x$; As x increases, y increases by a decreasing amount. Second differences are constant.
   b. missing values: (4, 28), (5, 30)
   c. The relationship is quadratic.

Parachuting Down, p. 29

1. slope = –700/2 or –350
2. rate of change = 350 meters per second downward
3. The domain for the line segment is $1 \le s \le 3$. The range for the line segment is $2{,}100 \le A \le 2{,}800$.

Sketching a Graph, p. 30

1.

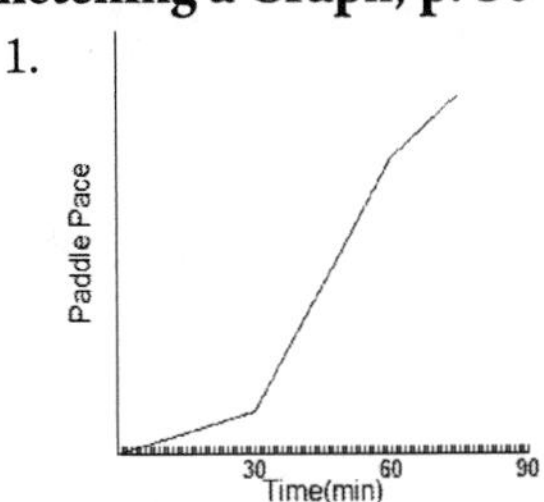

2.

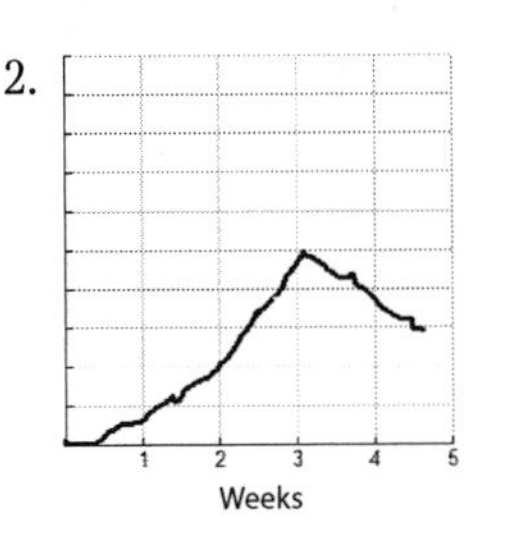

Interpreting Graphs I, p. 31

1. b
2. c

Interpreting Graphs II, p. 32

1. c
2. a

Ferris Wheel Graphs, p. 33

Students should choose to justify choice *c* as the proper graph.

Roller Coaster Ride, p. 34

1. between 4 and 6 seconds
2. about 13 m/sec; 48 m/sec
3. from 1 to 4 seconds into the ride; 6 to 7 seconds into the ride
4. The velocity increased rapidly. They stayed at 55 m/sec for 1 second, then decreased between 7 and 8 seconds, increased to about 53 m/sec, and then began to decrease again.

Graphing People Over Time, p. 35

Graphs will vary depending on students' interpretations of the situations. Most students will create line or bar graphs. Discuss the appropriateness of a continuous or discrete graph for these models.

Renting Canoes, p. 36

People	Total charge ($)
10	190
15	285
20	380
25	475
30	570
35	665
40	760
45	855
50	950

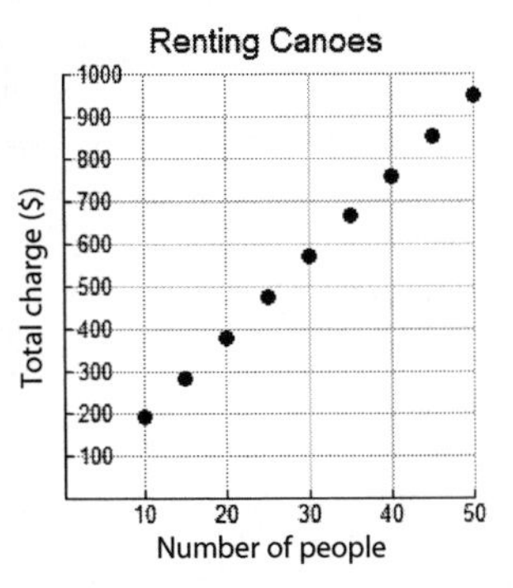

Grow Baby!, p. 37

The weight increases more each month in the first year than in the second year. The graph looks like it might fit a parabolic curve for $0 \le x \le 25$.

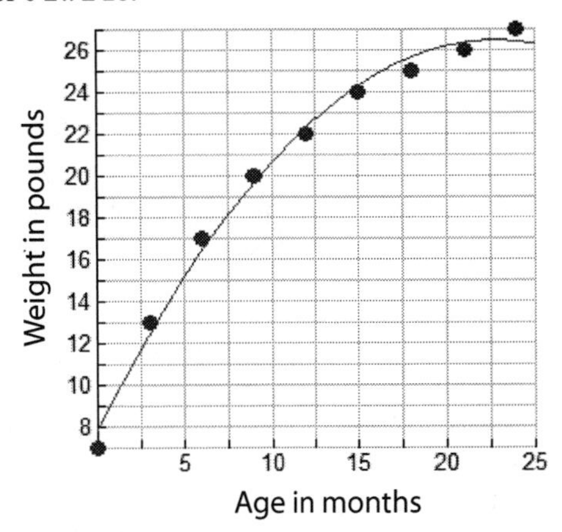

Quadratic regression
$\text{regEQ}(x) = -0.035955x^2 + 1.63514x + 7.82424$

Geometry

Double Trouble, p. 38

Students may decide that the triangles are both similar and congruent, or they may decide that they are congruent, but not similar. The correct answer is that they are both; however, students' answers are not important. What is important is that they are thinking about the question.

Up, Down, and All Around, p. 39

1. Flip it over (or flip it over the y-axis).
2. Move it up and to the right (or move it up 3 spaces and to the right 2 spaces).
3. Rotate it (or turn it). Then move down 5 spaces and right 1 space.
4. Make it larger and move it up and to the left (or make it two times larger and move it up 2 spaces and to the left 6 spaces).

Translate This!, p. 40

1. rectangular prism
2. sphere
3. triangular prism
4. circle

Mirror, Mirror on the Wall, p. 41

1. (–6, 1)
2. (6, 1)
3. (–3, 6)
4. (3, 6)
5. (–1, 3)
6. (1, 3)
7. The x-coordinates are opposite signs, but the same numbers. The y-coordinates are the same.

Copy Cat, p. 42

1. AB
2. BC
3. CA
4. yes
5. yes
6. yes
7. Sample response: Area of ABC is 1/4 the area of XYZ.

Tree Dilemma, p. 43

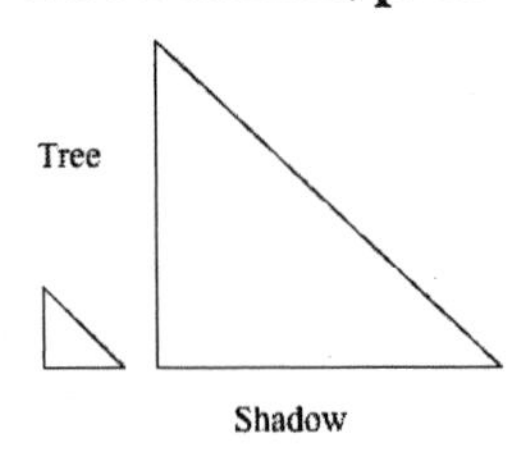

$\frac{47.25}{3.5} = \frac{T}{3}$, $T = 40.5$; Eric's garage may be safe, but only by a foot and a half.

Road Trip, p. 44

1.

Time in hours	Northbound car's distance in miles	Eastbound car's distance in miles	Distance between cars in miles
1	60	50	78.1
2	120	100	156.2
3	180	150	234.3
4	240	200	312.4

Each car's distance from the starting point is increasing by the amount of its rate of speed each hour. The distance between the cars is increasing by 78.1 miles each hour.

2. $100^2 - 80^2 = 3600 = 60^2$; thus, the other car has traveled 60 miles in 2 hours and is traveling at 30 mph.
3.

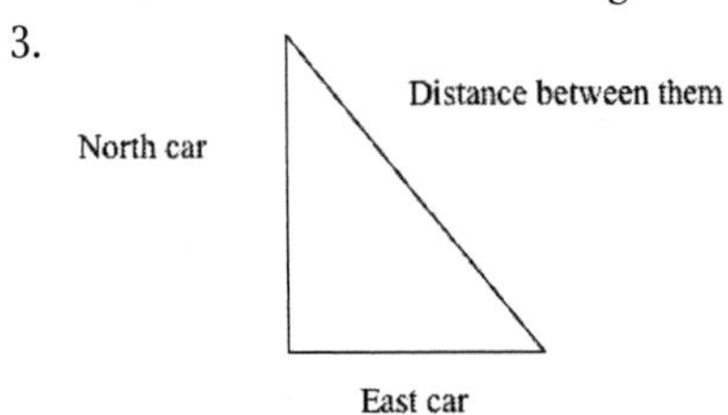

Comparing Cylinders, p. 45

1. 3:1
2. 3:1
3. 9:1
4. 27:1

Concession Concern, p. 46

1. The volume of the cone is ⅓ the volume of the cylinder.
2. $1.00

Popcorn Pricing, p. 47

Small container: $V = \pi r^2 h$; large container: $V = 4\pi r^2 h$.
The girls get twice as much popcorn in the large container as in two small containers.

Juice Packaging, p. 48

1. volume of prism = 187.5 cubic inches; volume of cylinder = 147.26 cubic inches
2. A comparable price for the prism container would be $4.19.

Cylinder Expressions, p. 49

b and c

The volume of a cylinder is given by the formula $V = \pi r^2 h$, where πr^2 = base area. The height, h, of the cylinder can be represented by $x + y$. The volume is then expressed as $V = 45(x + y) = 45x + 45y$.

Paper Cylinders, p. 50

1. Answers will vary.
2. volume of tall cylinder = 63.24 cubic inches; volume of short cylinder = 76.96 cubic inches; volume of tall prism = 49.67 cubic inches; volume of short prism = 60.5 cubic inches
3. The short cylinder is 16.46 cubic inches greater than the short prism.
4. Answers will vary.

Tennis Ball Packaging, p. 51

Volume of container = 509 cubic centimeters; volume of a tennis ball = 113 cubic centimeters; volume of air = 170 cubic centimeters. Answers will vary.

Statistics and Probability

Analyzing Scatter Plots, p. 52

Graph A looks like it might be linear. Graph B looks like a curve that opens upward; as x increases, y decreases. Graph C does not seem to have a best-fit line or curve.

Mr. Wiley's Baby, p. 53

1. linear regression ($a + bx$)
   regEQ(x) = 8.1875 + 0.458333x
2. Answers will vary.
3. Answers will vary.